QUANTUM THEORY

BULLET GUIDE

Hodder Education, 338 Euston Road, London NW1 3BH

Hodder Education is an Hachette UK company

First published in UK 2011 by Hodder Education

This edition published 2011

Artworks (internal and cover): Peter Lubach
Cover concept design: Two Associates

British Library Cataloguing in Publication Data: a catalogue record for this title is available from the British Library.

10 9 8 7 6 5 4 3 2 1

www.hoddereducation.co.uk

Typeset by Stephen Rowling/Springworks

Printed in Spain

QUANTUM THEORY

BULLET GUIDE

Jacob Dunningham

Acknowledgements

I am indebted to all my teachers, colleagues, students and friends at the Universities of Auckland, Oxford, and Leeds. Numerous discussions with them have answered my many questions and raised yet more but, above all, got me thinking. This book would not have been possible without the help of the staff at Hodder Education. I would particularly like to thank Harry Scoble and Sam Richardson for all their support and advice. Finally, I thank my family for being a constant source of encouragement and for giving perspective to all the theory.

This book is dedicated to my late father, John.

About the author

Jacob Dunningham is a Reader in Quantum Physics at the University of Leeds. Originally from New Zealand, he has studied at the Universities of Auckland and Oxford and now leads an active research team in quantum theory at Leeds. He is married with one son and, in a case of life imitating art, lives in a superposition of Leeds and London – an arrangement that will be ideal just as soon as the technology of quantum teleportation has been perfected. This is his second book on quantum physics.

Contents

Introduction

Quantum theory is one of the greatest intellectual triumphs in humanity's age-old quest for knowledge. It rules supreme in the **microscopic world**, describing the behaviour of the fundamental building blocks of our universe with astounding precision.

It has been tried and tested for more than a century without cracking, and its predictive powers are simply **unrivalled in all of science**. Despite this, many physicists have felt unsettled by the theory because it leads to some **bizarre and unexpected conclusions** about the nature of the universe.

Quantum theory assumes that quantities such as energy come in **discrete chunks** rather than having a continuous range of values. This makes the theory agree with what we see but opens up a mind-bending wonderland where things are not as they seem. Objects can be in **two places at once**, cats can be both **alive and dead** and particles can **pass straight through walls**. In this realm, we can no longer rely on our everyday intuition as a guide. Quantum theory requires a leap of faith but, so long as we are prepared to abandon our preconceived views of the world, it is surprisingly simple to apply and understand.

In this book, we give a brief overview of quantum theory and its remarkable consequences. We start at the very beginnings of its inception a little over a century ago, and develop through to some of its most exciting and cutting-edge applications. These include secret **codes that can never be cracked**, a whole new generation of **superfast computers** and the ability to **teleport objects** over large distances. These might seem like science fiction, but in the strange quantum world they are all very much science fact.

1 A new dawn

At the start of the twentieth century many people thought that physics was over. It seemed that there were just a few simple details left to iron out. Little did they know that physics was in for a big shock.

Explaining the niggling details needed a revolutionary new theory that completely overturned our ideas of nature and reality. That theory is **quantum physics** and it is the most accurate scientific theory we have.

Quantum physics is the most accurate scientific theory we have

Our future discoveries must be looked for in the sixth place of decimals.

Albert Michelson (1894)

In this chapter we will review how quantum physics was born in an almost desperate attempt to explain:

* blackbody radiation and the ultraviolet catastrophe
* the photoelectric effect
* Compton scattering.

In each case, the solution was to postulate that energy is not continuous, but comes in tiny chunks called **quanta** – from the Latin 'quantum' meaning 'how much?'. Quanta are so tiny we don't notice them in our everyday lives. However, they have a huge impact on the microscopic world.

Blackbody radiation

One of the great outstanding problems in physics in the nineteenth century was understanding the **spectrum** of a **blackbody.**

A blackbody is an object that absorbs all the radiation that falls on it. By contrast, a **whitebody** is like a mirror and reflects all the radiation. Many objects, such as the Earth and Sun, are good approximations to a blackbody.

A blackbody also emits radiation. If we imagine a closed box with a tiny hole in it, any light inside the box is continuously absorbed and re-emitted by the walls. The light that escapes from the hole must have been emitted rather than reflected and so is characteristic of the blackbody.

The frequency of blackbody radiation was known to depend strongly on the temperature of the box. In 1859, the German physicist, **Kirchhoff**, declared that understanding the blackbody spectrum was the holy grail of physics.

In 1792, Josiah Wedgwood, of ceramics fame, was the first to observe that you could judge the temperature of an object in a kiln from the colour of its glow. He also noted that, strangely, the colour didn't depend on what the object was made of.

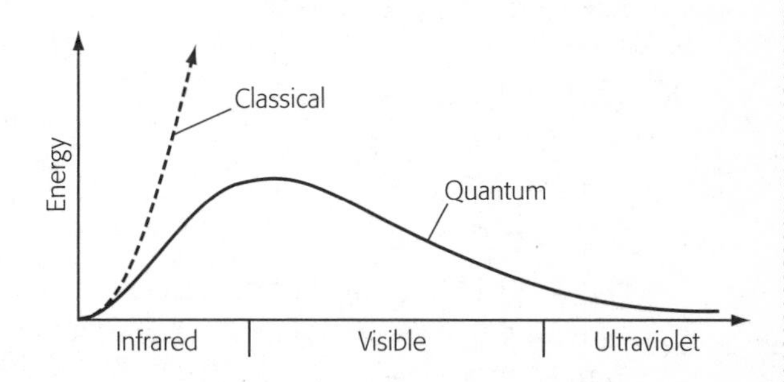

The energy of a blackbody for different colours of light. Classical physics predicts an infinite amount of energy for ultraviolet light but quantum physics resolves this problem

The ultraviolet catastrophe

Calculations based on classical physics predicted that a blackbody would radiate infinite energy at the ultraviolet end of the light spectrum. This contradicts what is observed, and is known as the **ultraviolet catastrophe.**

In 1900, **Max Planck** avoided this problem by taking the radical step of assuming that, rather than being continuous, the energy of light comes in discrete chunks called **quanta**.

The chunks that Planck postulated have an energy of $E = hf$, where f is the frequency of the light and h is a constant (now called **Planck's constant**) with a value of 6.63×10^{-34} joule seconds.

This postulate explained the experimental data perfectly and led to **Wien's law**, the relationship between temperature and wavelength:

$$\lambda = 2.9 \times 10^{-3}/T$$

where λ is the dominant wavelength (in metres) and T is the temperature (in Kelvin). The temperature in Kelvin is given by adding 273 to the temperature in Celsius.

Planck may have resolved the ultraviolet catastrophe but he left us with the uncomfortable conclusion that energy seems to be **quantized**.

The Sun appears yellow, i.e. it has a dominant wavelength of about 5×10^{-7} m. Wien's law allows us to calculate the surface temperature to be about 5800 K.

The photoelectric effect

In 1887, **Heinrich Hertz** observed another result that defied a classical explanation. He shone light on a metal and found that electrons were emitted. This is known as the **photoelectric effect** and has some curious features that couldn't be explained at the time:

- The speed of the emitted electrons was independent of the intensity of the light.
- The speed of the electrons instead increased with the *frequency* of the light.
- There was a critical frequency below which no electrons were emitted.

Classically, we would expect the electrons to gain more energy as the intensity (rather than the frequency) of the waves is increased.

Think of waves crashing on a beach. It is the tallest waves that have the most energy.

Einstein was the first to explain the effect. Like Planck, he assumed that light is quantized in chunks with energy hf and that there is some minimum energy, ϕ, needed to knock an electron out of the metal. This explains why the energy increases with frequency, and there is a minimum frequency for electrons to be emitted.

This can be summarized in the equation:

$$E_k = hf - \phi$$

where E_k is the kinetic energy of the emitted electron. No electrons are emitted when $hf < \phi$.

Light is quantized in chunks with energy *hf*

Einstein received his Nobel Prize for explaining the photoelectric effect and not relativity. He himself considered his paper on the photoelectric effect to be his only revolutionary contribution to physics.

Compton scattering

A third experiment that confounded classical physics was performed by **Arthur Compton** in 1923. He scattered X-rays from a carbon target and found that their wavelengths got longer – an effect now known as **Compton scattering**.

The change in the wavelength, $\Delta\lambda$, for a scattering angle, θ, was given by:

$$\Delta\lambda = h(1 - \cos\theta)/mc$$

where

- c is the speed of light
- m is the electron's mass
- h is Planck's constant.

This was puzzling as the wavelength shouldn't change.

Think of the electrons in the carbon like corks bobbing on water waves. They will bob at the same rate as the incident wave and so should always re-radiate light with the same wavelength.

Compton explained this result by assuming that light behaves as particles (called **photons**) rather than waves. His steps to deriving the **Compton formula** were:

1. Assume that the electron and photon can be treated like particles.
2. Take the momentum of the photon to be $p = h/\lambda$.
3. Apply both conservation of energy and conservation of momentum to the collision.

These three steps taken **together** give the Compton formula perfectly.

Along with the work of Planck and Einstein, Compton scattering gives compelling evidence that we should think of light as a collection of discrete quanta rather than a continuous wave.

2 Waves and particles

We normally think of light as a wave. However, in the last chapter we saw that we sometimes have to think of it as particles. So which is it? Is light a wave, a particle, or both?

Odd as it might seem, the answer depends very much on us. If we look for a wave, we see a wave. If we look for a particle, we see a particle.

**Is light a wave,
a particle,
or both?**

In this chapter we will explore this idea – called **wave–particle duality** – in more detail. We will show that it applies not just to light but to all objects and gives a good rule-of-thumb for whether we need to use quantum or classical physics to describe a system.

In particular, we will discuss:

* Young's **double-slit** experiment
* the effect of **measurements**
* De Broglie waves
* the **Davisson–Germer** experiment.

Wave–particle duality

Physicists use wave theory on Mondays, Wednesdays, and Fridays, and particle theory on Tuesdays, Thursdays, and Saturdays.

Henry Bragg

The debate over the nature of light dates back to the 1600s. Christiaan Huygens argued in favour of waves whereas Isaac Newton preferred a particle (or **corpuscle**) theory.

We have seen some arguments for why light must consist of particles, but there are equally compelling cases where only a wave theory will do.

If two beams of light overlap, they can give rise to a series of light and dark bands called an **interference pattern**.

1 **Constructive interference** occurs when the peaks of two waves coincide and they add up to give a bright spot.
2 **Destructive interference** occurs when the peaks of one wave coincide with the troughs of the other and they cancel out to give a dark spot.

The same effect can be seen with waves at a beach. Interference is a clear signature of waves; there is no way that a particle theory can describe this behaviour.

So light sometimes behaves like particles and at other times behaves like waves. This is known as **wave–particle duality**.

Double-slit experiment

Wave–particle duality is beautifully illustrated by the **double-slit experiment**, which was first performed by Thomas Young in 1801 – long before quantum physics was dreamed up.

He shone light at a barrier with two slits in it and a screen behind. If light was made of particles we would expect two bright spots behind the slits. Imagine someone throwing tennis balls at a wall with two holes in it. We would see tennis balls emerging behind the wall at two distinct places.

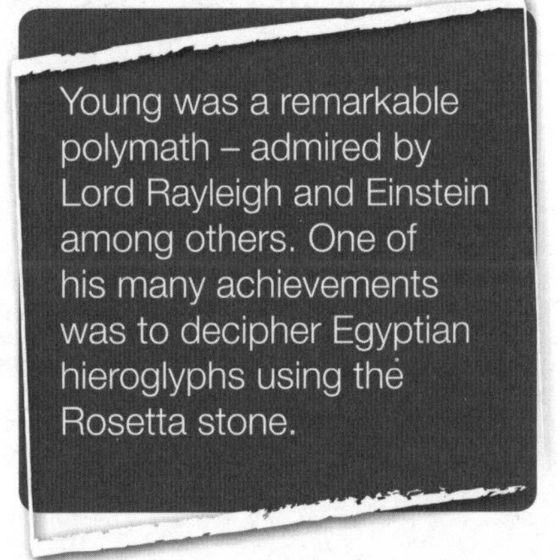

Instead he observed an **interference pattern** with lots of bright spots, indicating that light was behaving like a wave.

Strangely, if we were to reduce the intensity of the light until there was only one **photon** in the beam, we would still see interference. What is interfering with what? There is only one quantum of energy and it cannot be split in two.

The only explanation is that the photon passes through *both* slits at the same time and interferes with itself. This is called a superposition. The photon can be in two places at once!

The photon can be in two places at once

De Broglie wavelength

Things keep getting stranger. If we modify the experiment so that a detector records which slit each photon passes through, the interference pattern disappears and we see two spots corresponding to the two slits, i.e. we get the result for particles.

So light behaves like a wave or a particle depending on what we do. Nature is very obliging:

- If we look for a wave we see a wave.
- If we look for a particle we see a particle.

The double-slit experiment ... has in it the heart of quantum mechanics. In reality, it contains the only mystery ...

Richard Feynman

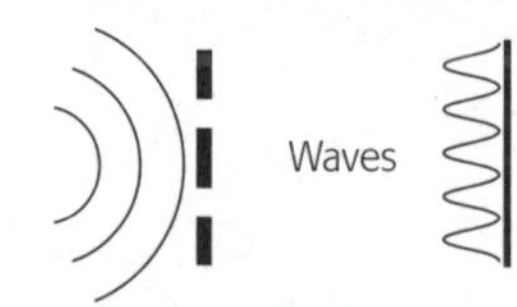

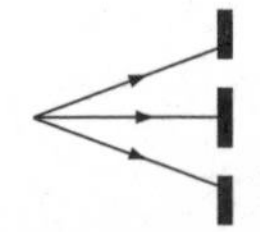

Particles

A French prince, **Louis de Broglie**, took the idea of wave–particle duality further. He argued that if light – which we normally think of as waves – can behave like particles, then matter – which we normally think of as particles – must sometimes behave as waves.

He wrote down a formula for the wavelength, λ, that he thought any object should have:

$$\lambda = h/p = h/mv$$

where

- p is the momentum of the object
- m is its mass
- v is its speed
- h is Planck's constant.

This is now known as the **de Broglie wavelength.** The de Broglie wavelength is longer for small, slow-moving objects.

Testing the theory

In 1927, **Davisson** and **Germer** tested de Broglie's theory. They fired electrons at a crystalline nickel target and measured the pattern of the scattered particles.

They observed that the electrons formed an interference pattern, proving that they were behaving like waves. The pattern was also consistent with the wavelength predicted by de Broglie, thus confirming his theory.

We can calculate the de Broglie wavelength for a human with mass 60 kg running at 5 m/s. This is about $\lambda = 2 \times 10^{-36}$ m – an unimaginably small distance. This is why we don't diffract around doorways. It seems true that in life we really don't make many waves!

Similar experiments have also seen wave-like behaviour for bigger objects. In 1999, **Anton Zeilinger** at the University of Vienna demonstrated interference for buckyballs – molecules that look like mini-footballs and consist of 60 carbon atoms. Experiments are currently under way to observe interference for even larger molecules.

TOP TIP

We can see whether an object behaves quantum mechanically by calculating its de Broglie wavelength and seeing whether it is similar to (or larger than) the objects it interacts with.

3 Inside the atom

The concept of a fundamental building block for all matter dates back to early Greek and Indian philosophers. The name of this unit – '**atom**' – comes from the Greek *'a tomos'*, meaning uncuttable or indivisible.

Experiments in the early twentieth century, however, revealed that atoms are not the last word. They can be divided and have an intriguing hidden structure. Quantum theory is crucial in describing and explaining this structure.

Atoms can be divided and have an intriguing hidden structure

It turns out that, to understand the behaviour of atoms, quantities other than energy must also be quantized. The universe is even more quantum than the work of Planck, Einstein and Compton might suggest.

In this chapter we will review the development of our current understanding of the atom. In particular, we will describe:

* Rutherford's scattering experiments
* the problems with Rutherford's model of the atom
* the solution proposed by **Bohr**
* the **spectrum** of light emitted from atoms.

Rutherford scattering

In 1909 **Ernest Rutherford** proposed an experiment that would radically change our view of the atom forever. He suggested to **Hans Geiger** and **Ernest Marsden** that they fire positively charged **alpha particles** at a sheet of gold foil and observe the scattering pattern.

It was known at the time that matter consisted of positive and negative charges, but it was thought that they were fairly evenly distributed. Rutherford therefore expected to see just some small deflection of the particles. He was in for a big shock.

It was almost as incredible as if you fired a 15-inch shell at a piece of tissue paper and it came back and hit you.

Ernest Rutherford

Sometimes the alpha particles bounced right back on themselves. This could happen only if the positively charged alpha particles were being repelled by highly concentrated regions of positive charge in the gold.

It was clear from this that the atom had some structure. Rutherford had 'split' the atom.

Rutherford's model of the atom

Rutherford's scattering experiment led to a new model of the atom. This consisted of a positively charged **nucleus** (which was later found to contain **protons** and **neutrons**) surrounded by a negatively charged cloud of **electrons**. The electrons were thought to orbit the nucleus much like planets orbit the Sun.

One way of picturing how incredibly small atoms are is that there are more atoms in a single grain of sand than grains of sand on all the beaches in the world! They are also mainly empty space: if the nucleus were the size of a golf ball, the nearest electron would be more than a kilometre away.

Unfortunately Rutherford's model is fatally flawed. An electron in orbit is accelerating because it is constantly changing its direction of travel.

The problem for the atom is that accelerating charges are known to emit energy, which means that the electrons should spiral into the nucleus. Classical physics says that atoms should be unstable and collapse in a tiny fraction of a second. This is not what we observe. So what is going wrong?

Classical physics says that atoms should be unstable and collapse in a tiny fraction of a second

Bohr's model of the atom

The Danish physicist **Niels Bohr** addressed the problem by saying that we can't treat atoms classically. They need to be described with quantum physics.

He didn't really resolve the problem with Rutherford's model – he just postulated the problem away! Much like Planck had done earlier, he justified his assumptions by the fact that they worked.

Bohr's postulates were:

1. Electrons move in **circular orbits** about the nucleus.
2. Electrons can orbit only at certain **discrete distances**.
3. Electrons in these allowed orbits **do not radiate energy**.
4. An electron can change orbits by absorbing or emitting light.

The angular momentum, L, for the electron is given by the formula $L = mvr$, where

- m is the electron's mass
- v is the speed of the electron
- r is the radius of the orbit.

This can be rewritten using de Broglie's wave formula, $mv = h/\lambda$, and the fact that an integer number of wavelengths must fit around the orbit, i.e. $2\pi r = n\lambda$, where n is an integer. This latter condition ensures that the wave matches up with itself at every point.

This gives $L = nh/2\pi$, and we see that angular momentum is quantized in units of $h/2\pi$, which is often denoted $\hbar$ and called the **reduced Planck constant**.

Atomic spectra

As well as having quantized angular momentum, each allowed orbit has a different discrete energy. According to Bohr's fourth postulate:

- electrons can jump to an orbit nearer the nucleus by emitting a photon or
- jump further away by absorbing a photon.

We know that white light (e.g. from the Sun) is a mixture of all colours because when it refracts through raindrops a continuous rainbow spectrum can be seen.

By contrast, light from an atom only gives discrete lines when passed through a prism.

The frequency (or colour) of the light that is emitted when an electron jumps between two orbits with an energy difference ΔE, is $f = \Delta E/h$.

A major triumph of Bohr's postulates was that they enabled him to derive a formula that predicts the wavelength of the **spectral lines** of hydrogen:

$$1/\lambda = R(1/n_f^2 - 1/n_i^2)$$

where

* n_f and n_i are integers denoting, respectively, the final and initial orbits involved in the transition. They have values 1, 2, 3 … starting at the innermost orbit.
* R is **Rydberg's constant** and has a value of about 1.097×10^7/m.

Many street lamps appear yellow rather than white. These create light from sodium atoms, which have dominant spectral lines in the yellow range. Sodium lamps cause less light pollution than ordinary lights and so are often used in cities with astronomical observatories.

4 The random universe

An unsettling consequence of quantum theory is that the universe is fundamentally random.

We normally think of physics as rigorous and deterministic. If we know the initial conditions of a system, we should be able to predict its future as well as determine all its past behaviour. This is not true in the quantum world. In fact, we cannot predict the outcome of even a simple experiment with certainty.

We cannot predict the outcome of even a simple experiment with certainty

This troubled Einstein greatly, and still troubles some physicists to this day. But things are not so bad since quantum theory at least gives a very precise theory of the *probability* of events.

In this chapter we will discuss:

* the inherently random nature of the universe
* the idea of a **wave function** and its links with probabilities
* the **Schrödinger equation**, which tells us how probabilities evolve.

I don't believe that God plays dice with the Universe.

Albert Einstein

Stop telling God what to do.

Niels Bohr

An unpredictable experiment

We can understand the inherent randomness of the universe with a simple experiment.

Suppose we have a sheet of glass that allows half of all light through and reflects the other half. This is called a 50:50 **beam splitter** and can be achieved by depositing just the right amount of silver on the surface.

Now suppose we have two detectors: one that clicks when a photon passes through the beam splitter and another that clicks when a photon is reflected.

If we fire a large number of photons at the beam splitter, we would expect about half to be recorded at each detector. We are unlikely to get exactly half, but something close. Just as, if we toss a coin 100 times, we are unlikely to get exactly 50 heads and 50 tails.

But what if we fire just one photon at the beam splitter? We know there is a 50:50 probability of it going each way, but we also know that the beam splitter cannot divide the quantum into two.

The only possibility is that the whole photon *randomly* goes to each detector half of the time. This is disturbing – it means that physics can't predict the outcome of this very simple experiment!

The wave function

Erwin Schrödinger gave probabilities in quantum physics a formal footing. Inspired by **wave–particle duality,** he came up with the idea of a probability wave. This is called a **wave function** and is usually represented by the Greek letter psi, Ψ.

However, it wasn't clear how the wave function should be interpreted:

Erwin with his Ψ can do
Calculations quite a few
But one thing that has not been seen
Just what does Ψ really mean.

Erich Hückel

Max Born resolved this issue in 1925 and was awarded the Nobel Prize in 1954 for doing so. His many contributions to physics have made him (in physics circles at least) even more famous than his granddaughter, Olivia Newton-John.

The Born interpretation

The modulus squared of the wave function (i.e. the wave function multiplied by its complex conjugate) is the probability density for a particle to be found at position x.

Born said that the probability of finding a particle between the two locations x_1 and x_2 is:

$$\int_{x_1}^{x_2} |\Psi(x)|^2 \, dx$$

Since the particle must be found somewhere and all probabilities must add up to 1, we have:

$$\int_{-\infty}^{\infty} |\Psi(x)|^2 \, dx = 1$$

This is the **normalization condition** – all wave functions must satisfy it.

The Schrödinger equation

The story goes that Schrödinger was once giving a talk about the wave nature of matter when someone in the audience asked, 'If matter really behaves like waves, then what is the wave equation governing its behaviour?'.

Schrödinger didn't know the answer, but immediately set about working it out. This resulted in the celebrated **Schrödinger equation**, which earned him the Nobel Prize in 1933.

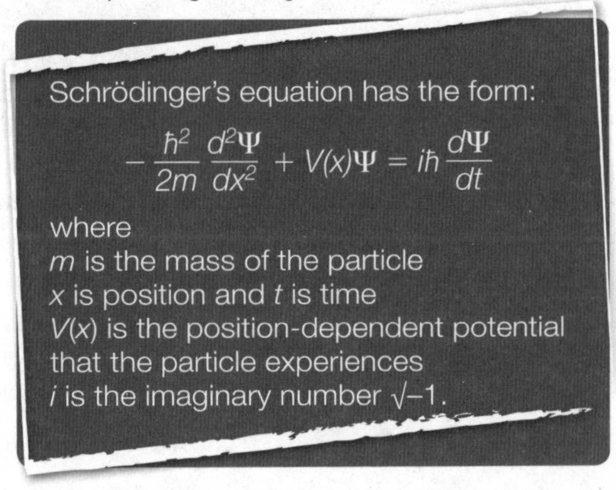

Schrödinger's equation has the form:

$$-\frac{\hbar^2}{2m}\frac{d^2\Psi}{dx^2} + V(x)\Psi = i\hbar\frac{d\Psi}{dt}$$

where
m is the mass of the particle
x is position and *t* is time
V(*x*) is the position-dependent potential that the particle experiences
i is the imaginary number $\sqrt{-1}$.

This equation cannot be derived from fundamental principles but is a *law* of physics. Once again, the reason we are happy to use it is simply that it works. No experiment has ever seen the slightest deviation from its predictions.

It is important because it tells us how quantum probabilities change with time and position. This is why quantum physics is often referred to as **quantum mechanics**.

The first term on the left of the equation represents kinetic energy and the second potential energy, and the term on the right is total energy. Schrödinger's equation is really just a quantum way of expressing the conservation of energy.

Schrödinger's equation is really just a quantum way of expressing the conservation of energy

Time-independent Schrödinger equation

In many cases, the total energy of a physical system is constant. This means that the right-hand side of Schrödinger's equation can be replaced with a constant energy, *E*, times the wave function.

Such a system is said to be in an **energy eigenstate** and the wave function is called an **energy eigenfunction.** This is just another way of saying that the total energy is fixed.

This simpler version of the Schrödinger equation is called the **time-independent Schrödinger equation** and has the form:

$$-\frac{\hbar^2}{2m}\frac{d^2\Psi}{dx^2} + V(x)\Psi = E\Psi$$

As the name suggests, time doesn't appear anywhere in this equation. This means that the solutions do not change with time. They are usually called **stationary solutions.**

To summarize:

1 At a fundamental level the universe seems to be **probabilistic.**
2 Probabilities are embodied in the **wave function** of a system.
3 **Born's rule** tells us that the modulus squared of the wave function gives the probability density for finding a particle at given location at some time.
4 The wave function evolves according to **Schrödinger's equation.**
5 When the total energy of a system is fixed, we can use the simpler **time-independent Schrödinger equation.**

5 Applying the theory

A large part of physics is concerned with making predictions about the future behaviour of systems. It is therefore worrying that, on the microscopic level, the universe seems to be fundamentally random.

Fortunately, Schrödinger and Born have given us a beautiful theory that determines the probabilities of certain events. So now we have a theory, but how do we use it?

Now we have a theory, but how do we use it?

In this chapter we will see how Schrödinger's equation can be applied to different situations. In particular, we will consider:

* the wave function of a particle in **free space**
* a particle trapped in an **infinite square well**
* a particle in a **quantum harmonic oscillator**
* **Heisenberg's uncertainty principle**.

These cases demonstrate how to apply the theory and, despite their simplicity, are often very good approximations to realistic systems.

A particle in free space

Let's start with the simplest case of all – a particle in free space. This means that there is no potential ($V = 0$) to trap the particle and so it is free to roam anywhere.

We can treat this system quantum mechanically by finding its wave function from the Schrödinger equation as follows:

1 Suppose that the particle has total energy E.
2 Substitute $V = 0$ into the time-independent Schrödinger equation, which gives

$$-\frac{\hbar^2}{2m}\frac{d^2\Psi}{dx^2} = E\Psi$$

3 Solve to get the two solutions $\exp(ikx)$ and $\exp(-ikx)$, where $k^2 = 2mE/\hbar^2$. These correspond to **plane waves.**
4 The general solution is any superposition (or sum) of these two solutions.

Now that we have the wave function, we can turn it into a more recognizable quantity.

Born told us that if we take the modulus squared of the wave function, we get the probability density for finding the particle at position x.

If we take the case of the general solution for a particle in free space, we get

$$P(x) = |\Psi|^2 = \text{constant}$$

The probability density does not depend on x. This means that the particle is equally likely to be found anywhere.

This is just what we might intuitively expect for a particle in free space. However, we have now managed to show this with a full quantum calculation.

The infinite square well

Next we consider an **infinite square well.** This consists of a potential that is zero over some range $0 \leq x \leq L$ and infinite otherwise.

Over the range $0 \leq x \leq L$, we have $V = 0$ and so the solution of the Schrödinger equation is the same as a free particle. Outside this range the wave function must be zero.

The potential is like an infinitely hard wall and so there is no chance that the particle can penetrate it.

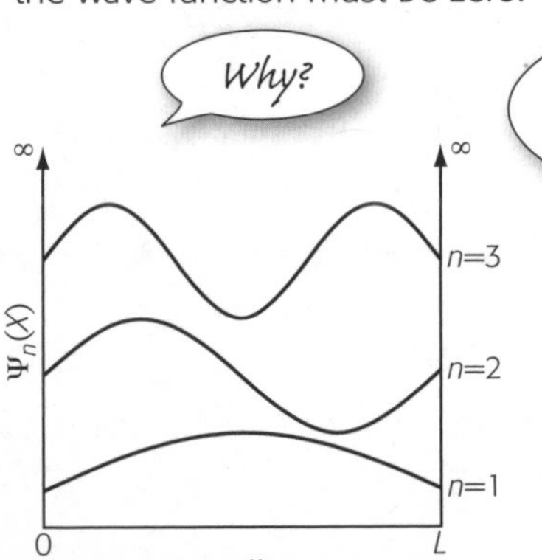

- Wave functions of the three lowest energy levels for the infinite square well

Imposing these **boundary conditions** gives us the general solution:

$$\Psi_n(x) = \sqrt{\frac{2}{L}} \sin\left(\frac{n\pi x}{L}\right)$$

where $n = 1, 2, 3, \ldots$ The first three of these are shown in the figure. These wave functions have corresponding energies

$$E_n = \frac{\pi^2\hbar^2}{2mL^2}n^2$$

which are quantized. This time, however, the quantization is not a postulate, but has emerged naturally from the Schrödinger equation.

We can see from the energy equation that, as we reduce the value of L, the energies increase. This is an example of **Heisenberg's uncertainty principle**, which we will meet shortly. It says that the more we confine a particle, the larger its average speed (and hence energy) becomes.

The quantum harmonic oscillator

Another important potential is the **quantum harmonic oscillator.**

This is just the quantum version of a marble rolling back and forth in a parabolic bowl. Its potential has the form

$V(x) = 2\pi^2 m f^2 x^2$ where f is the frequency of oscillation.

The wave functions of the three lowest energies are shown in the figure and have energies:

$En = (n + 1/2)hf$

where $n = 0, 1, 2 \ldots$ Again the energies are quantized, but now they have equal spacing $\Delta E = hf$. This is precisely the quantization that Planck assumed.

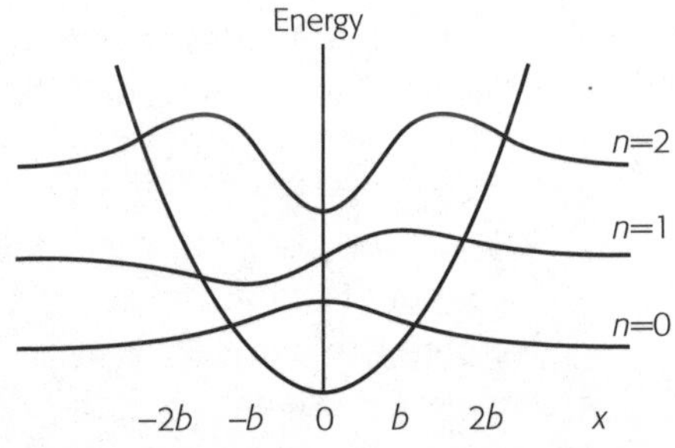

Wave functions of the three lowest energy levels for the quantum harmonic oscillator

The value b in the figure corresponds to the **classical turning point** of the lowest energy state. This is the furthest distance that a marble with energy $E = hf/2$ could roll up the walls of the bowl before it stops and rolls back down.

Intriguingly, the wave function of the lowest energy state extends beyond b. A quantum particle can roll further up the walls of the bowl than is possible classically!

If we solve the Schrödinger equation for an electron trapped in the potential of an atom we recover the quantization of energy and angular momentum that Bohr postulated. It seems that his guesses were right after all.

The uncertainty principle

In 1927, **Werner Heisenberg** showed that the more certain we are about a particle's position, the less certain we are about its momentum.

This is known as **Heisenberg's uncertainty principle** and can be written as $\Delta x \Delta p \geq \hbar/2$, where Δx and Δp are the uncertainties in position and momentum. Since their product cannot be smaller than $\hbar/2$, as one goes down the other must increase.

This means we can never know both the position and momentum of an object precisely.

An inscription on Heisenberg's grave reads: 'He lies here, somewhere'.

We can never know both the position and momentum of an object precisely

The more precisely the position is determined, the less precisely the momentum is known in this instant, and vice versa.

Werner Heisenberg

The uncertainty principle is a consequence of wave–particle duality. It doesn't make sense to think of the precise location of a wave on a string. Similarly, if we send a localized pulse down the string, the wavelength and hence the momentum, $p = h/\lambda$, is not well defined. As either the wave's position or its momentum gets sharper, the other becomes less so.

The uncertainty principle is an important feature of quantum physics that distinguishes it from classical theories.

6 Tunnelling

Now we have seen how to apply quantum theory, we can use it to unveil some startling features of the physical world. One such feature is called **quantum tunnelling**, which is what occurs when quantum particles pass straight through barriers and emerge on the other side. This seems incredible, as classical physics forbids it, but is a very real effect.

Quantum particles can pass straight through potential barriers and emerge on the other side

Imagine throwing tennis balls at a brick wall. It wouldn't matter how many balls you threw, you would expect them all to bounce back at you. That is because they behave classically and never have enough energy to pass through the wall. The same is not true in the quantum world.

Quantum tunnelling explains certain types of radioactive decay and has even been used in new technologies.

In this chapter we will discuss:

- quantum tunnelling through a square barrier
- how tunnelling naturally occurs in alpha decay
- the application of tunnelling in scanning tunnelling microscopes.

The classically forbidden region

When we studied a harmonic potential in the previous chapter, we saw that a quantum particle could sometimes roll further up the side of a bowl than was possible classically.

It is said that the particle can enter the **classically forbidden region.** In other words, a quantum version of a tennis ball would sometimes be found *inside* the wall. We can understand this behaviour by solving the time-independent Schrödinger equation for the case $E < V$, i.e. when the particle doesn't have enough energy to get over the barrier.

The solutions are $\Psi(x) = \exp(\pm\alpha x)$, where $\alpha^2 = 2m(V - E)/\hbar^2$, and physically the wave function tends to decay exponentially inside the barrier.

This means that the probability for finding the particle in the **classically forbidden region** decreases rapidly the further into it we go.

However, if the barrier is thin enough, there is some chance that the particle will reach the other side and 'pop out'. In other words, it can successfully tunnel through the barrier.

We can understand this by considering the case of a **square barrier**.

A square barrier

A square barrier consists of a potential that has a constant value, V, over some region $0 \leq x \leq L$, and is zero everywhere else.

If we consider a free particle that is incident on the barrier from the left and has energy, $E < V$, we can find the probability of tunnelling by:

1. Solving Schrödinger's equation in the three regions, i.e. to the left, right and inside the barrier.
2. Matching the wave functions at the two boundaries.
3. Calculating the ratio of the input wave function on the left to the output wave function on the right.

The resulting probability of tunnelling is:

$$T = \left(1 + \frac{V^2 \sinh(\alpha L)}{4E(V-E)}\right)^{-1}$$

A useful approximate formula for a general barrier is:

$$T \approx \exp\left(-\frac{2}{h}\int \sqrt{2M(V(x)-E)dx}\right)$$

where the integral is over the classically forbidden region. This shows that the rate of tunnelling decreases exponentially as the barrier gets higher and wider.

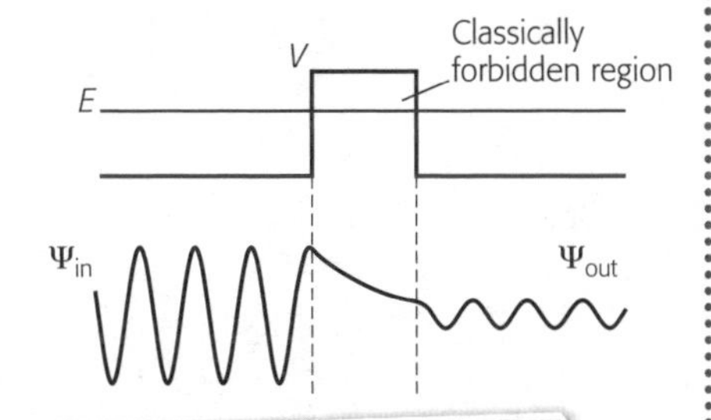

Tunneling of a particle with energy *E* through a square barrier with height *V*

> Suppose a 50 kg man walked at 1 m/s into a constant 30 J barrier that was 1 m wide. His kinetic energy would be 25 J and his probability of tunnelling would be about exp(–4.3 × 10^{35}) – an incredibly small number. This is why humans tend not to pass through walls.

Alpha decay

One important process that can be explained by tunnelling is the decay of radioactive nuclei by the emission of alpha particles. We have met alpha particles before in the scattering experiments conducted by Rutherford to determine the structure of atoms. They are charged particles consisting of two protons and two neutrons.

In the early days of quantum mechanics, the alpha decay of radioactive nuclei was a perplexing problem: nuclear forces were seen to be too strong for any particle to be able to leave the nucleus. In fact, that is the whole reason why we have nuclei at all. They hold together because nuclear forces are far stronger than electrostatic repulsion.

What's more:

* Alpha particles emitted from different sources all have similar energies.
* However, the rates of decay vary by many orders of magnitude.

George Gamow was the first to explain alpha decay in 1928. He considered an alpha particle inside the nucleus with insufficient energy to overcome the strong nuclear potential. However, with some small probability, the alpha particle could tunnel through the barrier and escape.

We have seen that the tunnelling probability depends exponentially on the difference between the particle's energy, E, and the barrier height $V(x)$. This explains why small variations in the energies can lead to very large differences in the rates of emission.

Scanning tunnelling microscope

In 1981, **Gerd Binnig** and **Heinrich Rohrer**, two scientists at IBM in Zürich, came up with a new technology that revolutionized the field of microscopy. They invented the **scanning tunnelling microscope** (STM) – a device that uses quantum tunnelling to image surfaces on the atomic scale. This breakthrough earned them the 1986 Nobel Prize in Physics.

An STM works by scanning a conducting tip over a material very close to the surface. A voltage is applied between the tip and the surface and electrons will tunnel between the two, resulting in a weak electric current. The size of this current depends exponentially on the distance between the tip and the surface.

By monitoring the current as the tip moves over the surface, it is possible to resolve features on the surface with incredible resolution. STMs allow us to image and manipulate individual atoms. Their resolution is more than 2000 times better than the best optical microscopes.

STMs are widely used in both industrial and fundamental research. Their uses include:

- characterizing surface roughness
- observing surface defects
- determining the features of molecules and aggregates on the surface.

STMs allow us to image and manipulate individual atoms

7 Scaling things up

We have now covered the quantum theory of individual particles. However, to understand the behaviour of larger real-world objects, we need a multiparticle theory.

It is tempting to think that we can just apply the individual particle theory separately to all the particles. After all, if we know the behaviour of each particle, we must know the behaviour of the whole system.

To understand the behaviour of larger real-world objects, we need a multiparticle theory

While this idea is largely true, there are some subtleties in how we need to combine particles in the quantum world.

In this chapter we will describe how this is done. In particular, we will discuss:

* bosons and fermions
* the Pauli exclusion principle
* the periodic table
* molecular bonding.

When quantum theory is scaled up to larger systems, its true power becomes clear. It underpins all of chemistry and hence, it could be claimed, is the theoretical basis of biology and life itself.

Bosons and fermions

Quantum particles are **indistinguishable,** which means we can never tell apart two particles of the same type.

If we consider two identical particles, swapping them shouldn't change anything measurable. Basic logic also dictates that, if we swap them twice, we must get back to where we started.

There are only two ways these can both be true. Swapping the particles once either:

1. Doesn't change the wave function. Particles of this type are called **bosons**.
2. Or gives the wave function an overall minus sign: $\Psi \rightarrow -\Psi$. This minus sign is not measurable since, as Born told us, we only measure $|\Psi|^2$. Particles of this type are called **fermions.**

These two cases behave very differently.

Fermions tend to avoid each other whereas bosons like to clump together.

All particles of a given type have an intrinsic angular momentum of fixed magnitude $S = \hbar\sqrt{[s(s + 1)]}$, where s is called the **spin** and can take the values $s = 0, 1/2, 1, 3/2, 2, \ldots$

- Bosons are particles with integer s values.
- Fermions are particles with half-integer s values.

A curious feature of spin-half particles, such as electrons, is that we have to rotate them twice (720 degrees) to get back to their starting point!

Fermions tend to avoid each other whereas bosons like to clump together

The Pauli exclusion principle

The symmetry properties of fermions lead to a very important principle, first identified by Wolfgang Pauli in 1925.

This is called **Pauli's exclusion principle** and can be stated as:

> No two identical fermions can occupy the same space simultaneously.

We can intuitively understand this as follows. If two identical fermions are swapped, the wave function must be unchanged. However, since they are fermions we also know that we must get the negative of the wave function.

The only way that these can both be true is if the wave function is zero, i.e. there are no particles. So all fermions have to be different.

This simple idea has far-reaching consequences including:

* the physical properties of solids
* the stability of large objects against collapse
* an understanding of white dwarfs and neutron stars in astrophysics
* the electronic structure of atoms.

Electrons are called spin-half particles, which means that they have $s = 1/2$ and so $S = \hbar\sqrt{3}/2$. The direction of the intrinsic angular momentum, S, is quantized and can take two values called spin-up and spin-down. This means that at most two electrons can be at a given location and they must have opposite spins.

The periodic table

In 1869, the Russian chemist Dmitri Mendeleev arranged the elements that were known at the time into a table based on their chemical properties. This is known as the **periodic table** and is a very useful tool for understanding the structure of atoms and their behaviours.

The story goes that the periodic table came to Mendeleev in a dream. He wrote it down when he woke and, amazingly, only one correction had to be made later.

The size of the atomic weight determines the nature of the elements.

Dmitri Mendeleev (1869)

The periodic table can be understood by solving Schrödinger's equation for an electron in an atom. This gives different energy levels called **orbitals,** which, in order of ascending energy, are: 1s < 2s < 2p < 3s < 3p < 4s < 3d < 4p. Electrons sequentially fill these levels subject to the fact that the s-orbitals can have two electrons, the p-orbitals six and the d-orbitals 10.

How can some orbitals have more than two electrons? Doesn't this violate the exclusion principle?

No. The letters label different orbital angular momenta and have sublevels corresponding to different directions. Each sublevel can contain up to two electrons with opposite spins.

Molecular bonding

More complex chemical and biological systems are created when atoms bond to form molecules.

The main types of **molecular bonding** are:

* ionic bonding and
* covalent bonding.

Ionic bonds form when two oppositely charged ions are bound by electrostatic attraction. This happens between elements from opposite sides of the periodic table.

For example, a sodium (Na) atom can lose an electron to become Na^+ and a chlorine (Cl) atom can gain an electron to become Cl^-. These then attract one another to form sodium chloride (NaCl).

It costs energy to create the ions, but if the energy gained by the attraction outweighs this, bonding is energetically favourable.

What about bonding between atoms of the same type such as in hydrogen (H_2) molecules?

In this case, it is never energetically favourable for ionic bonds to form and a different mechanism – called **covalent bonding** – takes place.

Covalent bonding involves the two atoms sharing their electrons. The wave functions of the electrons on each atom can either add symmetrically ($\Psi_1 + \Psi_2$) or antisymmetrically ($\Psi_1 - \Psi_2$).

The symmetrical combination is energetically favourable. It has a higher density of electrons between the two nuclei than the antisymmetrical wave function. This concentration of negative charge acts as a 'glue' that binds the two positive nuclei together.

8 Quantum paradoxes

Quantum theory applies to situations that are very different from our everyday experience. Its effects are most keenly felt for objects that have large de Broglie wavelengths and so are extremely small and cold.

This means that the predictions of quantum theory often defy our common sense. If we try and use our intuition to explain the quantum world, we are often confronted with **paradoxes**.

The predictions of quantum theory often defy our common sense

These paradoxes are not real because there is nothing inconsistent about them. They arise from our preconceived views about the way the world should behave.

Nonetheless, they have provoked a lot of debate about the nature of reality, and have often been used to try and pick holes in quantum theory.

In this chapter we will review some of the more well-known paradoxes. In particular, we will discuss:

* the measurement problem
* the quantum Zeno effect
* Schrödinger's cat
* Wigner's friend.

I think I can safely say that nobody understands quantum mechanics.

Richard Feynman

The measurement problem

No phenomenon is a physical phenomenon until it is an observed phenomenon.

John Wheeler

Quantum theory has a split personality:

* If left alone, a quantum system can be in a superposition of different quantities such as position or momentum.
* However, if we measure the system, we only ever see a single outcome.

In Young's double slits, for example, we know that each particle goes through *both* slits. Yet detectors always detect each particle at one or the other. The detection process is said to **collapse the wave function** to one of its possible values.

This strange dichotomy between the behaviour of observed and unobserved systems is known as the **measurement problem.**

Along with measurements, systems can also lose their quantumness simply by interacting with the environment. This process is called **decoherence** and is used to explain the boundary between the quantum and classical worlds. Everyday objects constantly interact with the environment, which is why we don't see them behaving quantum mechanically.

The quantum Zeno effect

An intriguing consequence of quantum measurement is the **quantum Zeno effect**. This is based on the arrow paradox proposed by the ancient Greek philosopher Zeno of Alea, which states that, since an arrow in flight does not move in any instant of time, it cannot possibly move at all.

The quantum version of this paradox shows that motion is impossible for a continuously monitored system. It's really a quantum demonstration of the old adage that 'a watched kettle never boils'.

Suppose we measured the position of a quantum particle and then let it evolve before measuring again. The second measurement is likely to reveal a different position because the particle has moved in the meantime.

Now suppose we measured it many times in quick succession. Since the system can't evolve much between measurements, there is a high probability that each measurement will collapse the wave function back to the original position. In other words, the motion is frozen.

The Zeno effect has been experimentally verified in unstable quantum systems. By continually measuring them, they were prevented from decaying. In different measurement regimes, it is also possible to measure an **anti-Zeno effect**, where the motion is *enhanced* by measurement.

Schrödinger's cat

One of the most famous paradoxes arising from quantum theory, was put forward by Schrödinger in 1935 in an essay on the conceptual problems with quantum theory. This is now known as **Schrödinger's cat.**

Imagine the following scenario:

- A cat is placed in a sealed box.
- In the box is a phial of deadly gas.
- The phial is connected to a radioactive nucleus.
- If the nucleus decays, it triggers the release of the gas.

Now suppose we wait until there is a 50% chance that the nucleus has decayed. Because it is a quantum particle, the nucleus will then be in a superposition of decayed and not decayed.

Crucially, the state of the nucleus is perfectly correlated with whether the cat is alive or dead. If the nucleus decays, the cat is dead, and, if it doesn't, the cat is alive. This means that the cat must be in a superposition of being alive and dead at the same time.

However, if we were to open the box and check on the unfortunate cat, we would always find it either alive or dead – never both. So is the cat really in a superposition before we look, and, if so, how does looking destroy it?

When I hear about Schrödinger's cat, I reach for my gun.

Stephen Hawking

Wigner's friend

This raises the question of what constitutes a measurement. A person looking inside the box clearly does, but what about a speck of dust interacting with the cat, or a flea making the observation?

These considerations led **Eugene Wigner** to propose a related thought experiment to illustrate his belief that **consciousness** is important in the measurement process. This is known as **Wigner's friend.**

Imagine that a friend of Wigner performs the Schrödinger cat experiment after Wigner leaves the laboratory. Only when he returns does Wigner discover the outcome. The question is: was the superposition of 'sad friend/dead cat' and 'happy friend/live cat' destroyed only when Wigner returned, or some time earlier?

Wigner argued that the collapse happened as soon as the friend opened the box and that it was the friend's consciousness that *caused* the collapse.

But, should consciousness really play a role in quantum theory? How complex does a life form have to be to have consciousness, and how drunk do we have to be to not?

The current prevailing view is that consciousness is not important. Either **decoherence** destroys the superposition before the box is opened. Or we can adopt the **many worlds interpretation** of quantum mechanics (see Chapter 10), which avoids the need for a collapse at all.

Should consciousness really play a role in quantum theory?

9 Entanglement and non-locality

One of the most perplexing features of quantum theory was identified by Schrödinger in 1935. He called it *Verschränkung* and it is now known by its English translation **'entanglement'**.

Entanglement is a strange type of connectedness that can occur between quantum systems. If we do something to one particle in an entangled pair, it can instantaneously affect the other one, even if they are separated by a large distance.

Entanglement is a strange type of connectedness that can occur between quantum systems

In other words, the system can behave **non-locally**: something here can affect something over there. This conclusion greatly troubled Einstein, and he was moved to call entanglement 'spooky action at a distance'.

Despite his misgivings, entanglement is a real feature of the quantum world and is even exploited in a range of new technologies.

In this chapter we will introduce the ideas of entanglement and non-locality and discuss some applications. We will look at:

* Bell's inequality and hidden variables
* quantum teleportation
* quantum computing
* quantum cryptography.

Entanglement

We are used to **correlations** in the classical world: lack of sleep is correlated with tiredness and summer is correlated with warmer weather. However, unlike the classical world, quantum theory allows superpositions of different correlations.

Imagine, for example, we had two spin-half fermions that could each either be spin-up or spin-down. Now suppose there was some interaction between them that ensured that their spins always pointed in the same direction, i.e. they were perfectly correlated. One possible state is:

$$|\uparrow\rangle_1|\uparrow\rangle_2$$

We have written the state using **Dirac notation**. This is just a shorthand way of saying that both particle 1 and particle 2 have spins pointing up. The other state is:

$$|\downarrow\rangle_1|\downarrow\rangle_2$$

Quantum mechanics allows the system to be in the superposition:

$$|\uparrow\rangle_1|\uparrow\rangle_2 + |\downarrow\rangle_1|\downarrow\rangle_2$$

This is an *entangled state*, because we cannot write the total state as the state of one particle multiplied by the state of the other. This is a useful definition of entanglement.

If the spin of one particle is measured to be pointing up, then the other particle will instantly point up also. This is true even if the particles are separated by a long distance. This strange 'action at a distance' is called **non-locality**.

Einstein, Podolsky and Rosen could not stomach non-locality and, in 1935, they claimed that it proved that quantum theory was incomplete.

Bell's inequalities

This led to a debate over the existence of so-called **hidden variables.** These are variables that do not appear in the formalism of quantum theory (hence hidden), but predetermine the outcomes of measurements.

If true, this idea would avoid the uncomfortable notion of non-locality because the measurement of one particle would no longer affect the other, distant, particle.

In 1964, the Northern Irish physicist **John Bell** found a way of distinguishing the two possibilities. He imagined that:

- Two distant parties (Alice and Bob) each receive one of the particles.
- Alice and Bob each randomly choose one of two measurements to make.
- The process is repeated many times.

Bell showed that, if he combined the outcomes of Alice and Bob's measurements in a particular way, he arrived at an interesting inequality.

This provided a bound that local hidden variable theories must obey. Quantum theory, however, could violate this bound. This famous result is known as **Bell's inequality** and allowed experiments to test non-locality for the first time.

Numerous experiments have subsequently been performed and shown violations of Bell's inequality. There are still some **loopholes** in these experiments so the case is not completely closed. However, the evidence to date is overwhelmingly in favour of non-local quantum theory and against hidden variables.

Quantum teleportation

An intriguing application of non-locality is **quantum teleportation**.

We all know about teleportation from *Star Trek*, where a person is dematerialized and reappears on the surface of an alien planet. In quantum teleportation, a **quantum state** is instantaneously transferred to a distant particle.

The steps are:

1. Alice and Bob share an entangled pair of particles.
2. Alice makes a joint measurement of the particle she wishes to teleport and her particle from the entangled pair.
3. Alice phones Bob to tell him the measurement outcome.
4. Bob adjusts his particle based on this outcome.

Alice's measurement destroys her state and something similar to it instantly appears at Bob's end.

The phone call is crucial to let Bob know what he has to do to make the final state identical to the starting one. Teleportation has been achieved in experiments with photons, ions and atoms.

Doesn't teleportation violate relativity, which says nothing can travel faster than light?

Are you happy to accept a reverse charge call from the Starship Enterprise?

No. Relativity says no information can travel faster than light and no information is transferred by teleportation until the phone call is made. This last step, of course, is limited by the speed of light.

Quantum computers and cryptography

Another application is **quantum cryptography**, which allows completely secure communication between distant parties.

A secure cipher is established by Alice creating quantum states and sending them to Bob, who measures them. Any would-be eavesdropper is immediately identified because his measurements unavoidably alter the quantum state, changing Alice and Bob's results.

If an eavesdropper is detected, Alice and Bob discard their cipher and wait until a secure channel can be established.

Two can keep a secret if one is dead.

Anon.

Quantum cryptography is a mature technology that has even been used to collate election results in Switzerland and to coordinate security at the 2010 FIFA World Cup in South Africa.

Quantum computers are able to perform certain tasks exponentially faster than conventional computers

Quantum theory is also exploited in computation. **Quantum computers** are able to perform certain tasks exponentially faster than conventional computers.

The basic premise is that a quantum system can be in a superposition and so can represent many numbers simultaneously. This means that performing a single operation on a quantum system can be equivalent to many operations using a classical one. We get parallel computing for free.

Quantum computers are known to be particularly efficient at searching unsorted lists and factorizing large numbers. The latter is important because it is the basis of a lot of modern data transmission security.

10 What does it all mean?

Quantum theory is an amazing tool. It allows us to make predictions about the universe with astounding precision.

However, for all the questions quantum theory answers, it raises yet more. Many physicists believe that quantum theory is more than a calculational tool and reveals deep truths about nature.

A number of different **interpretations of quantum theory** have been put forward that try to understand the meaning behind the theory.

For all the questions quantum theory answers, it raises yet more

These can provoke intense debate among their most ardent supporters. Though at the moment we have no way of distinguishing between them. Any or none of them could be true. Which, if any, you subscribe to is really a matter of taste.

> **The opposite of a profound truth may well be another profound truth.**
>
> Niels Bohr

In this chapter we will introduce some of the main interpretations of quantum theory. These include:

* the Copenhagen interpretation
* the de Broglie–Bohm theory
* the many worlds interpretation
* collapse theories
* the 'shut up and calculate' approach.

The Copenhagen interpretation

The **Copenhagen interpretation** was developed in the 1920s by a number of people including Bohr and Heisenberg. It is named after Bohr's home town, where much of the seminal work was carried out.

It attempts to explain what quantum theory really means. Because it was a developed by a number of different people, there are different (sometimes contradictory) versions of it. We have been implicitly using the Copenhagen interpretation in the book so far.

Its key principles are:

- A system is described by a **wave function.**
- Nature is probabilistic with probabilities given by **Born's rule.**
- Measuring devices are classical objects and cause the wave function to **collapse**.

I like to think that the moon is there even if I don't look.

Albert Einstein

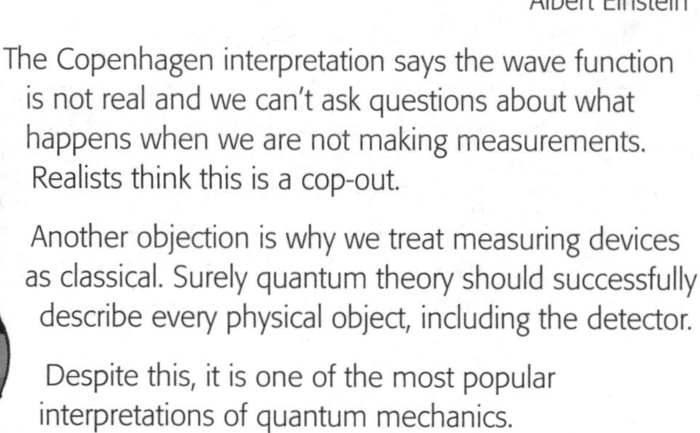

The Copenhagen interpretation says the wave function is not real and we can't ask questions about what happens when we are not making measurements. Realists think this is a cop-out.

Another objection is why we treat measuring devices as classical. Surely quantum theory should successfully describe every physical object, including the detector.

Despite this, it is one of the most popular interpretations of quantum mechanics.

The de Broglie–Bohm theory

An alternative interpretation was put forward by Louis de Broglie and later extended by David Bohm. This is known as the **de Broglie–Bohm theory** or sometimes **pilot wave theory.**

This interpretation attempts to resolve the **measurement problem** as well as the mystery of **wave–particle duality**. It says that:

- Particles always have well-defined positions.
- These particles are guided by a pilot wave.
- The pilot wave obeys the Schrodinger equation.
- The motion of the particles is **deterministic.**

Wave–particle duality is embraced by the de Broglie–Bohm theory by having a probability wave that governs the location of the particles.

It differs from the Copenhagen interpretation in that the particles always have a particular configuration even if no one observes them. This avoids the whole measurement problem because there is no need to postulate a collapse of the wave function.

It is also a **deterministic** theory in the sense that the same initial probability wave and configuration of particles will always give the same measurement outcome.

The predictions of the de Broglie–Bohm theory are the same as the Copenhagen interpretation and so experiments can't decide between them.

The many worlds interpretation

The **many worlds interpretation** was developed by **Hugh Everett III** in 1957. It has risen in prominence in recent years and is now favoured by many researchers working in the field.

This interpretation reconciles the measurement problem in quite a different way. It says that the universe is constantly splitting: every possible outcome of an experiment occurs, each one in a parallel universe. This means that our universe is embedded in a much larger **multiverse.**

In the Schrödinger cat experiment, for example, in one universe the nucleus decays, and the cat is dead and in another no decay occurs and the cat is alive. The observation doesn't affect reality.

The many worlds interpretation is cheap on assumptions but expensive on universes.

Paul Davies

This approach avoids issues such as the **collapse of the wave function** and **Wigner's friend.** A bizarre consequence is that there are many copies of us in other universes, each with a different history. Everything that could have happened to us in the past, but didn't, has happened to a copy of ourselves in a parallel universe!

Everything that could have happened to us in the past, but didn't, has happened to a copy of ourselves in a parallel universe!

Everett believed in quantum immortality – the idea that if you die in one universe you live on in a parallel one. He asked that when he died his ashes be thrown out with the household rubbish. His wife duly obliged.

Shut up and calculate

Another idea is that there is some mechanism that causes the wave function to collapse. There are different variants of this idea, which are collectively called **collapse theories.**

Perhaps the best known is the **Ghirardi–Rimini–Weber (GRW) theory**, which says that the wave function spontaneously collapses.

Roger Penrose has proposed that the superposition of an object in two different positions will collapse due to the gravitational field between them. This gives a limit to how large an object can be before quantum theory fails. Experiments are being built to test this idea, but have been inconclusive so far.

It's difficult to make predictions, particularly about the future.

Niels Bohr

Before we get carried away, there is a much more pragmatic approach. It's called **'shut up and calculate'.** This simply says that physics does an amazing job at calculating the behaviour of the physical universe, but we shouldn't push it too far. We should stop worrying about what it all means and just get on with calculating.

On that salutary note, perhaps it's time for me to end our whistlestop tour of quantum theory and get back to my own calculations.

Further reading

Quantum Theory Cannot Hurt You: A Guide to the Universe by Marcus Chown (Faber and Faber, 2008).

The God Effect by Brian Clegg (St Martin's Griffin, 2009).

Introductory Quantum Mechanics and Relativity by Jacob Dunningham and Vlatko Vedral (Imperial College Press, 2010).

In Search of Schrödinger's Cat by John Gribbin (Black Swan, 1985).

Physical Principles of the Quantum Theory by Werner Heisenberg (Dover, 2003).

The Bluffer's Guide to the Quantum Universe by Jack Klaff (Oval, 1997).

Quantum: Einstein, Bohr, and the Great Debate about the Nature of Reality by Manjit Kumar (Icon, 2010).

Introducing Quantum Theory by J.P. McEvoy and Oscar Zarate (Icon, 2007).

Quantum Bits and Quantum Secrets by Oliver Morsch (Wiley VCH, 2008).

Quantum Physics: A Beginner's Guide by Alastair I.M. Rae (Oneworld Publications, 2005).

Quantum Physics: Illusion or Reality? by Alastair I.M. Rae (Canto, 2004).

Decoding Reality by Vlatko Vedral (Oxford University Press, 2010).